OnBoard
ACADEMICS

Energy, Force and Motion

© 2015 OnBoard Academics, Inc
Portsmouth, NH
800-596-3175
www.onboardacademics.com
ISBN: 978-1-63096-046-9

OnBoard Academic's books are specifically designed to be used as printed workbooks or as on-screen instruction. Each page offers focused exercises and students quickly master topics with enough proficiency to move on to the next level.

OnBoard Academic's lessons are used in over 25,000 classrooms to rave reviews. Our lessons are aligned to the most recent governmental standards and are updated from time to time as standards change. Correlation documents are located on our website. Our lessons are created, edited and evaluated by educators to ensure top quality and real life success.

Interactive lessons for digital whiteboards, mobile devices, and PCs are available at www.onboardacademics.com. These interactive lessons make great additions to our books.

You can always reach us at customerservice@onboardacademics.com.

Forces and Motion

www.onboardacademics.com

A force is a push or a pull.

Place a √ in the box to indicate if the motion is a push or a pull.

	push	pull
kick		
slide		
drop		
throw		
roll		

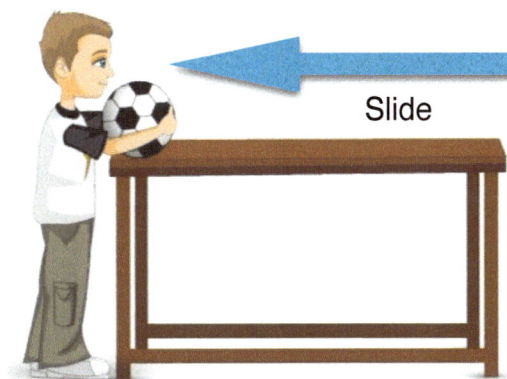

Slide

Roll

The Strength of the Force and the Size of the Object

The more massive an object is, the greater the force that is needed to make it move.

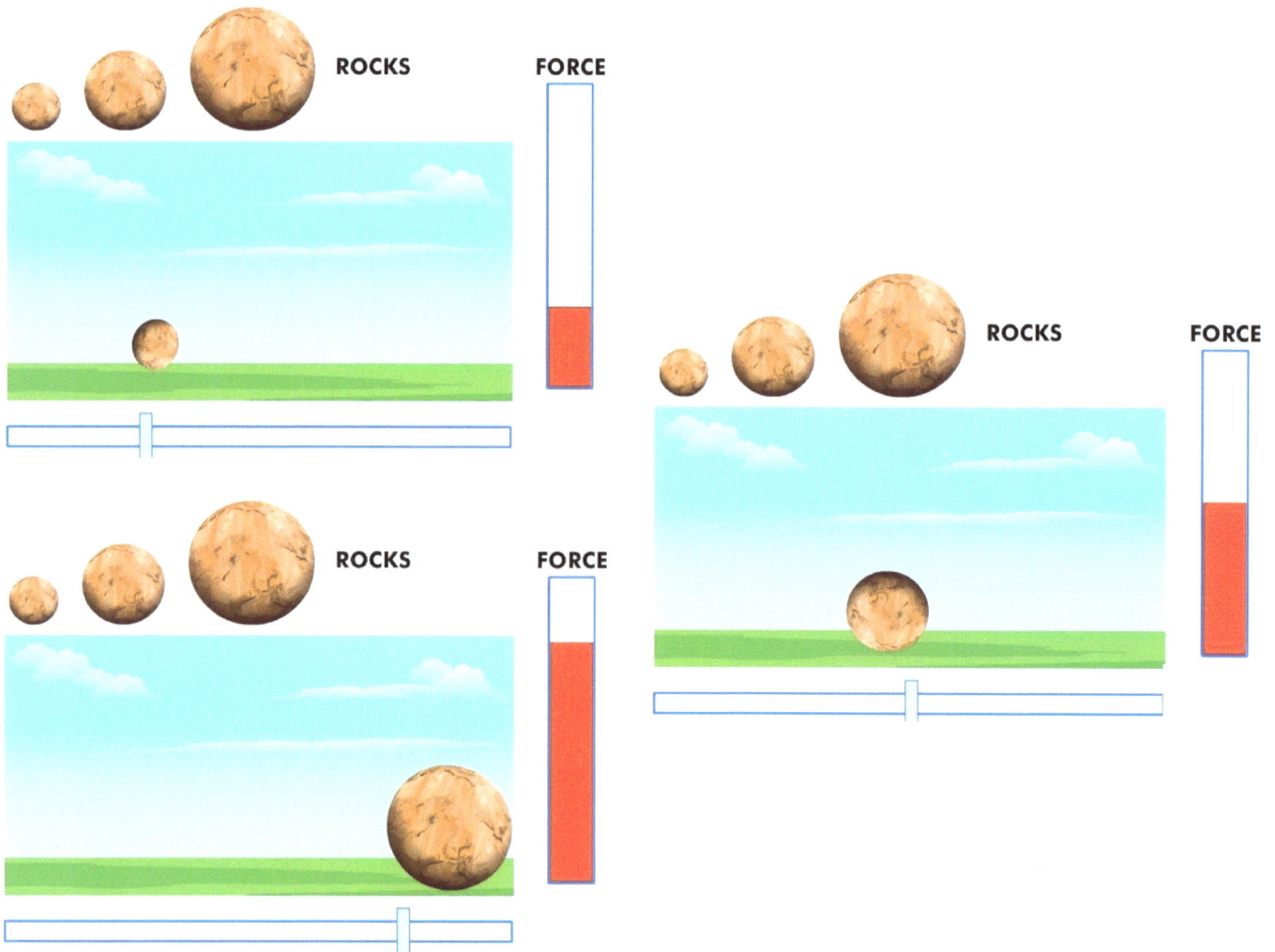

ROCKS   FORCE

ROCKS   FORCE

ROCKS   FORCE

A larger force will move an object farther and faster than a smaller force. With the same amount of force, a smaller, lighter object will move farther and faster than a heavier one.

Which force and mass combinations will make these outcomes.

Either draw the answer in the box provided or connect the empty box with the correct illustration.

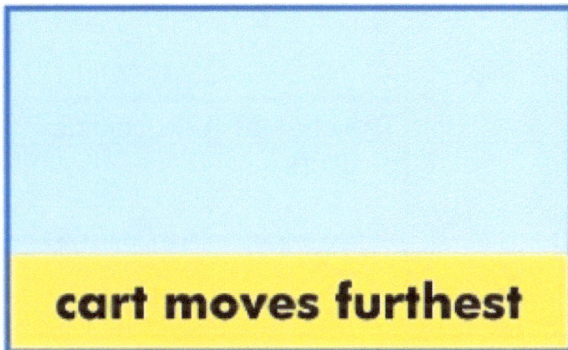

cart moves furthest	**cart moves least**

Force and Motion Review

Connect the vocabulary word with the correct definition.

mass	The location of an object.
force	The change in an objects position.
motion	A push or a pull that can change an objects motion.
speed	A measure of how fast an object moves.
friction	A force that slows the motion of an object.
position	If this is greater, more force is need to make this object move.

www.onboardacademics.com

We don't always notice motion.

Some things move so quickly we don't notice that they are moving. Light moves at about 300,000 km per second. That's fast enough to circle the Earth seven times in one second.

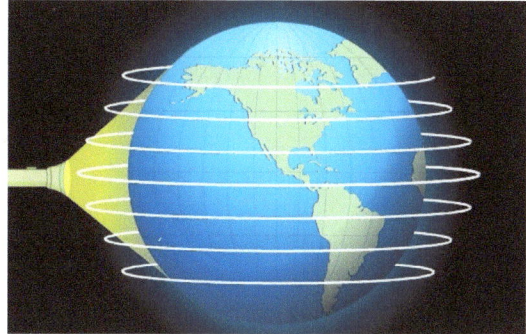

The Earth also travels very quickly. At the equator, the Earth is spinning about 1,700 km per hours. But the rotational speed seems like a snail's pace when you consider that the Earth orbits the sun at about 107,000 km per hour.

If we are rotating this quickly and traveling around the sun so fast, why don't we notice?

We don't notice because everything around us is moving at exactly the same speed. Think about flying in an airplane. You don't realize you are moving that fast because everything around you including the plane is also moving at exactly the same speed.

On the other hand, some things are moving so slowly you don't notice them moving either. Earth's continents are moving at 2.5 centimeters per year. That might not sound like much but in a million years that's 25 km and a million years isn't very long in geological terms.

Forces and Motion Quiz

1. _____ always has a direction, and is caused by force.
    a. Work
    b. Energy
    c. Motion
    d. Velocity

2. _____ is a force that slows down moving objects.
    a. Acceleration
    b. Motion
    c. Velocity
    d. Friction

3. The _____ an object, the _____ the force required to move the object.
    a. lighter, greater
    b. heavier, lesser
    c. heavier, greater

4. When a large force is applied, an object moves a _____.
    a. smaller distance
    b. larger distance

Energy and Work

What are energy and work?

Energy causes things to happen all around us. In science we say that energy has the ability to cause motion or to create change. For example, the energy we get from food helps us to grow and to do things like to kick a soccer ball.

When we put gasoline in the car we are using the gasoline to create chemical energy to create motion energy to turn the wheels of the car.

When we are at home and turn on a light, we are converting electrical energy into light energy.

When we kick a ball, drive a car or turn on a light switch, we say that energy is doing work. In science, work is defined as the transfer of energy that occurs when an object is moved a distance, or when something undergoes a chemical change.

 www.onboardacademics.com

So when the boy pushes this table and the table moves in the direction in which the boy is pushing the table we would say that work has been done.

If the boy pushes against this giant rock and the rock doesn't move, we would say that no work has been done even though the boy has used a lot of energy and feels as though he has done a lot of work.

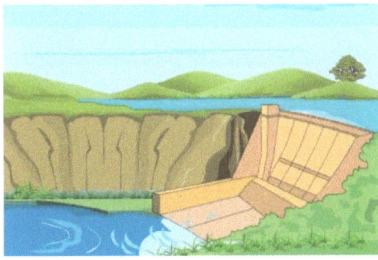

You can find energy and work in all sorts of places. The water in this dam has a type of energy that we call potential energy. Potential energy is stored energy. When the dam opens the stored potential energy is transferred into motion energy. Because the water has moved, we say that work has been done.

There are many different forms of energy including chemical energy, motion energy, heat energy and electrical energy.

Most of the energy on Earth comes from the sun. Energy from the sun warms the Earth, gives us daylight that is used in the process of photosynthesis. This is the process of plants making their own food. Sunlight is converted into chemical energy within the leaves of a plant.

Connect the correct energy source to each scene to enable work to be done.
Describe what work is being done in each scene.

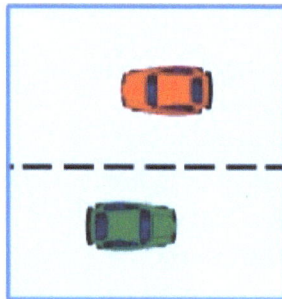

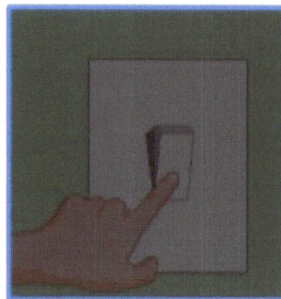

What work is being done?	What work is being done?	What work is being done?	What work is being done?

Label the different forms of energy.

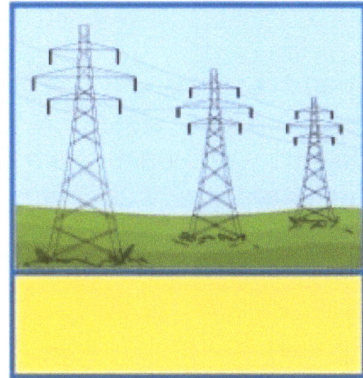

motion   wind   heat

light   electrical   chemical

When work is being done, energy often changes from one form into another.

When we switch on a light we are using electric energy. The electrical energy is transferred into light energy.

Let's take a look at how electricity is produced and reaches our homes. Along the way, we'll see how energy is transferred into different forms as work is being done.

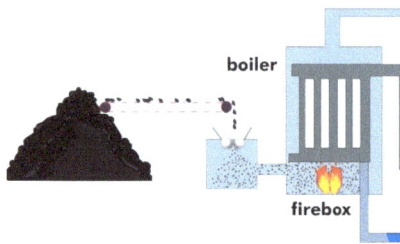

Electric energy usually starts with coal. Coal is used in half the power plants within the USA.

When coal is burned in a power plant it heats water in a boiler to very high temperatures creating a lot of steam. At this point work is done because water is heated and turned into steam and energy is transferred because chemical energy from the coal is transferred into heat energy.

The steam is piped from the boiler at very high pressure and used to turn large blades called turbines. At this point work is being done because the turbines turn and energy is transferred because the heat energy from the steam is transferred to motion energy.

The turbine is connected to a generator. The generator has large magnets that are surrounded by coils of copper

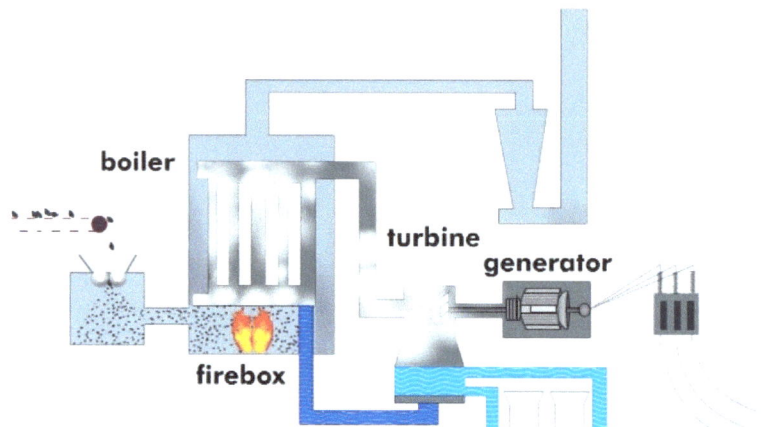

wires. The turbine turns the magnets causing them to spin within the copper coils and electricity is created. Work is done as magnets spin and energy is transferred because the motion energy from the turbine is transferred to electrical energy.

     www.onboardacademics.com

Electricity reaches our home on transmission wires. These are the wires you see on the

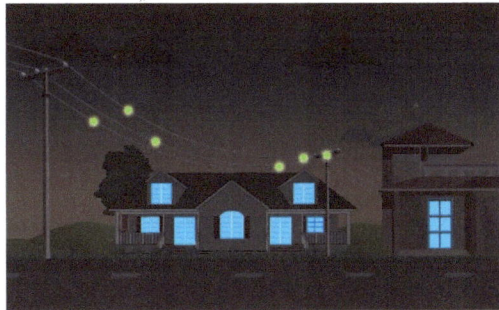

tall poles on the street. Sometimes these wires run underground and are hidden from your view.

Once the energy reaches our home we might transfer it into may different types of energy including heat to cook our food, sound energy to listen to our music or motion energy to run on a treadmill. Whatever the case work is being done!

List 5 types of work done by energy sources in your home.

1. _____

2. _____

3. _____

4. _____

5. _____

Which forms of energy are being transferred in the illustrations below?

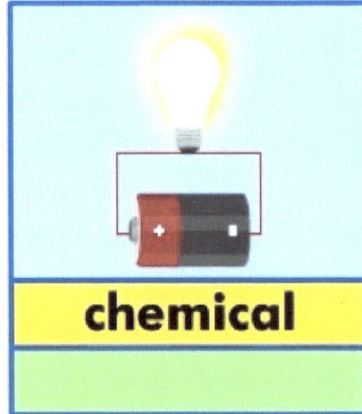

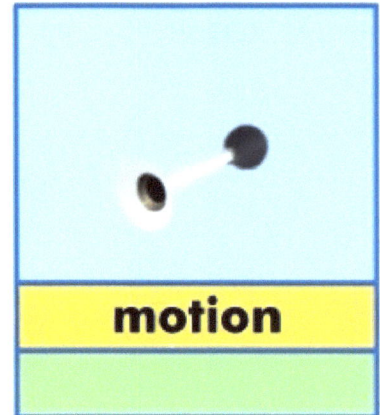

chemical	chemical	heat
motion	electrical	motion

sound   electrical   electrical
     wind   motion   chemical

Can you find and circle eight different types of energy?

```
A  X  M  O  T  I  O  N  C  D
Z  F  B  K  F  K  G  U  N  T
A  C  O  W  T  S  I  C  R  F
L  H  S  I  I  O  S  L  E  E
I  E  B  G  T  N  U  E  E  H
G  M  B  D  P  N  D  A  W  E
H  I  O  L  O  D  Z  R  G  A
T  C  E  G  T  R  S  R  I  T
N  A  T  S  U  N  D  O  R  J
E  L  E  C  T  R  I  C  A  L
```

Match!

potential	I'm the energy found in food, gasoline and batteries.
the Sun	I'm the energy that causes music and noise.
sound	I'm stored energy not yet in motion.
chemical	I'm the energy generated at a power station.
work	I'm the source of most energy on Earth.
electrical	This is done when an object moves a distance or a substance undergoes a chemical change.

Energy and Work Quiz

1. Wind is a form of energy. True or false?

2. The faster something moves, the _____ energy it has.
   a. more
   b. less

3. Oil, gas and coal all contain stored _____ energy.
   a. electric
   b. light
   c. heat
   d. chemical

4. A stove uses _____ energy to turn water into steam.
   a. wind
   b. motion
   c. heat
   d. chemical

5. Work is done when a mass is moved a distance. True or false?

6. The energy found in our food is in the form of chemical energy. True or false?

Changing Motion

www.onboardacademics.com

Which direction do the following objects move?
Fill in the blanks.

back & forth

around

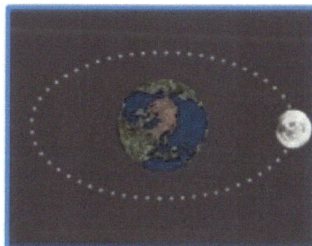

up & down

side to side

When things move, we say they are in motion. Motion always has a direction (for example, side to side, up and down, back and forth, or around), and we know something is in motion because it changes its location.

     www.onboardacademics.com

Pushes and Pulls Change Motion
Look at each action and decide if it is a push or a pull and label it push or pull.

A push or a pull is a force, and forces can change the way that things move. For example, a push or a pull can make an object start moving or make a moving object change its direction or slow down.

push

pull

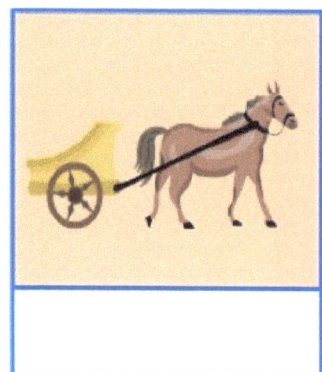

Can you guess the speed of these animals? Draw them on the number line according to their speed.

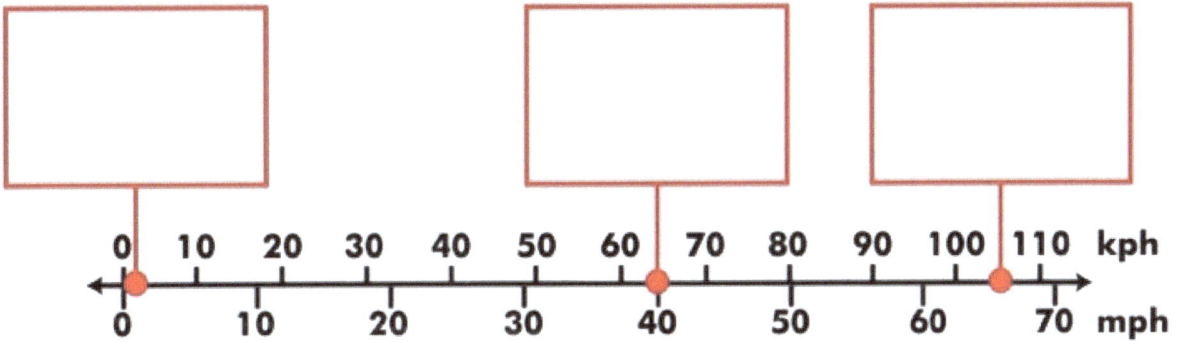

| 0 | 10 | 20 | 30 | 40 | 50 | 60 | 70 | 80 | 90 | 100 | 110 | kph |
| 0 | | 10 | | 20 | | 30 | | 40 | | 50 | | 60 | | 70 | mph |

Speed measures how fast an object is moving. kph and mph means kilometers or miles per hour: how many kilometers/miles the animal could run in an hour if it could maintain its speed.

Will this ball ever stop rolling? _____

Friction is a force that slows down moving objects. If the ball was rolling on a bumpy surface, it would slow down more quickly because there would be more friction. If the ball was rolling on a smooth surface, it would take longer to slow down because there would be less friction.

www.onboardacademics.com

Connect the definition with the correct word.

The location of an object	**force**
A change in an objects position	**motion**
A push or a pull that can change an objects motion	**speed**
A measure of how fast an object moves	**friction**
A force that slows the motion of an object	**position**

www.onboardacademics.com

Changing Motion Quiz

1. An object in motion has _____.
    a. Direction
    b. Speed
    c. Both direction and speed

2. Force is:
    a. Motion
    b. Push
    c. Pull
    d. Push or pull

3. A force that can slow down a moving object is called
    _____.
    a. Velocity
    b. Friction
    c. Density

4. Identify if this force is a push or a pull. _____.

Ways an Object will Move

Label each item with the way in which it will move.

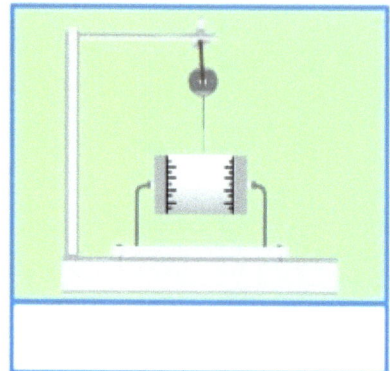

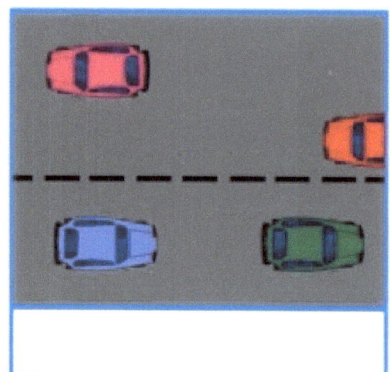

round and round   straight line   side to side

back and forth   zig zag   up and down

www.onboardacademics.com

Pushes and Pulls Change Motion

Label each action push or pull.

push

pull

A push or a pull is a force, and forces can change the way that things move. For example, a push or a pull can make an object start moving or make a moving object change its direction or slow down.

How are these objects moving?

Read the descriptions and then label each item by the way in which it moves.

	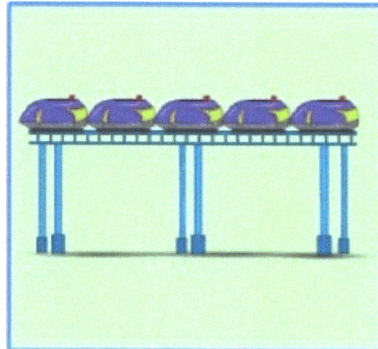	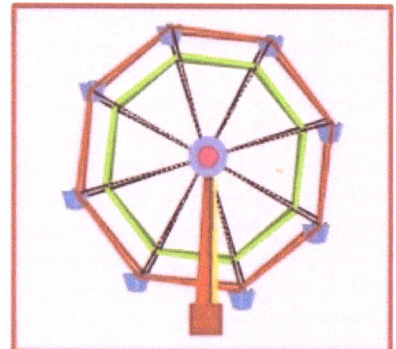
How and Why?	**How and Why?**	**How and Why?**

Motion that goes round and round happens because the middle point is fixed. In this Ferris wheel ride, the force comes from the center.

Motion that goes in a straight line happens because there is a force that moves it in one direction.

Motion that goes back and forth (or **side to side**) happens because there is a force that moves it in one direction and a force that moves in back in the opposite direction.

Will this ball ever stop rolling? Why? _____

_____

_____

Friction is a force that slows down moving objects. If the ball was rolling on a bumpy surface, it would slow down more quickly because there would be more friction. If the ball was rolling on a smooth surface, it would take longer to slow down because there would be less friction.

Which way will the ball move and by how much?

Use the red arrows to indicate the direction and the amount the know will move. Please note that some red arrows are longer than others indicating more movement.

Ways an Object will Move Quiz

1. _____ is a result of force.
   - a. Velocity
   - b. Motion
   - c. Revolution
   - d. Tension

2. The motion that goes _____ happens because the middle point is fixed.
   - a. back and forth
   - b. round and round
   - c. up and down
   - d. side to side

3. Motion that goes _____ happens because there is a force that moves the object in one direction.
   - a. up and down
   - b. side to side
   - c. in a straight line
   - d. in a zig zag pattern

4. Friction is a force that pushes in the direction that an object is moving, causing the motion to speed up. True or false?

Transferring Energy

www.onboardacademics.com

Potential and Kinetic Energy

There are six main forms of energy and many different types within each form but all energy exists in one of two states; potential energy and kinetic energy.

The energy stored in the water that is being blocked by the dam is potential energy. When the dam opens and the water is flowing downstream this energy become kinetic energy.

Potential energy is stored energy while kinetic energy is energy in motion.

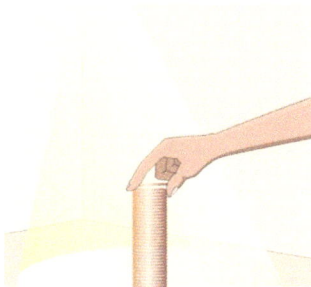

There are many different types of potential and kinetic energy. For example, a coiled spring has potential energy but when the spring is released it become kinetic energy.

A giant rock on top of a hill has potential energy but when it rolls down the hill the potential energy is transferred to kinetic energy.

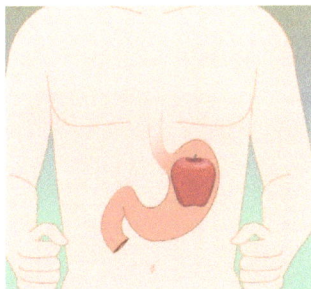

The food that is stored in our body as chemical energy is potential energy but it is converted to kinetic energy every time that we move.

There are six main *forms* of energy, (and many different types of energy within each form), but energy exists in two states: potential energy and kinetic energy. Potential energy is stored energy, while kinetic energy is energy in motion.

Potential energy can change into kinetic energy and vice versa.

Examine the rolling ball.

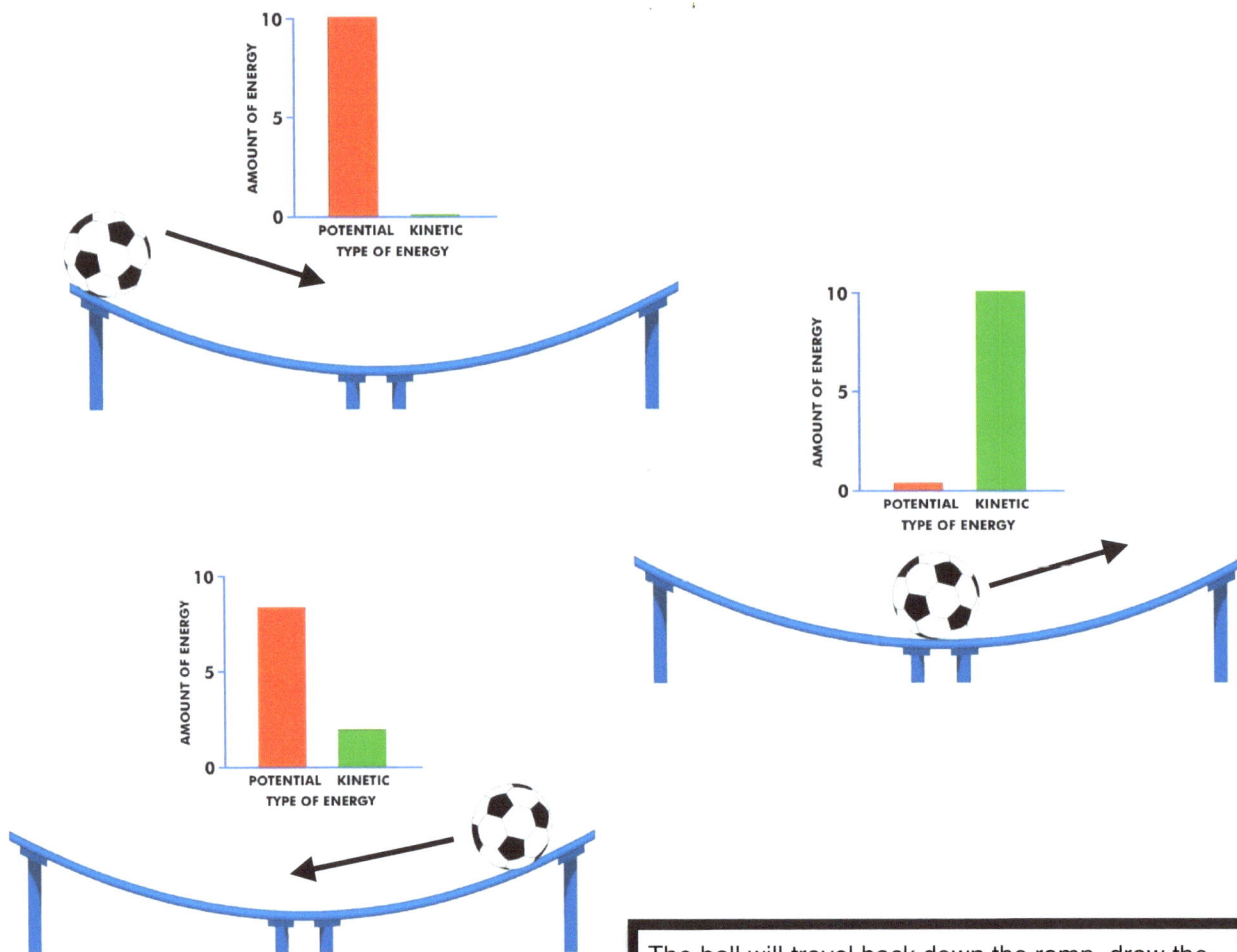

The ball will travel back down the ramp, draw the graph to represent the energy levels when it is at the bottom of the ramp on its way back up.

www.onboardacademics.com

Identify the energy status on the swing set.

1	Kinetic energy is being transferred into potential energy
2	Potential energy is being transferred into kinetic energy
3	Potential energy is zero; kinetic energy is at its maximum
4	Kinetic energy is zero; potential energy is at its maximum

Friction and Heat Energy

Will this ball ever stop rolling

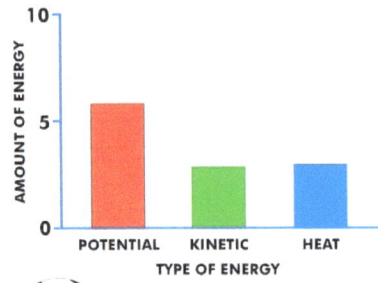

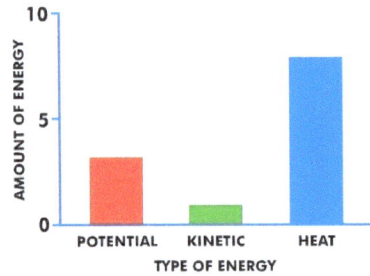

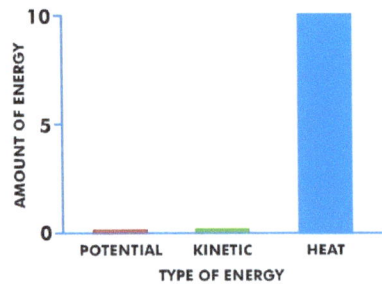

The soccer ball will eventually stop moving because there is **friction** between the ball and the ramp. When the ball rubs against the ramp, some of the ball's kinetic energy is transferred into **heat energy** in the ramp. The ball slows down, and the ramp heats up a little bit.

Identify the 6 forms of energy.

◯	◯	◯
The energy stored inside atoms	The energy of waves	Potential energy stored in the bonds between atoms

Ⓒ **chemical**   Ⓣ **thermal**   Ⓦ **wave**

Ⓜ **mechanical**   Ⓝ **nuclear**   Ⓔ **electrical**

◯	◯	◯
The energy of heat	The energy of electrical charges	The energy of the motion or position of an object

Organize this energy web.

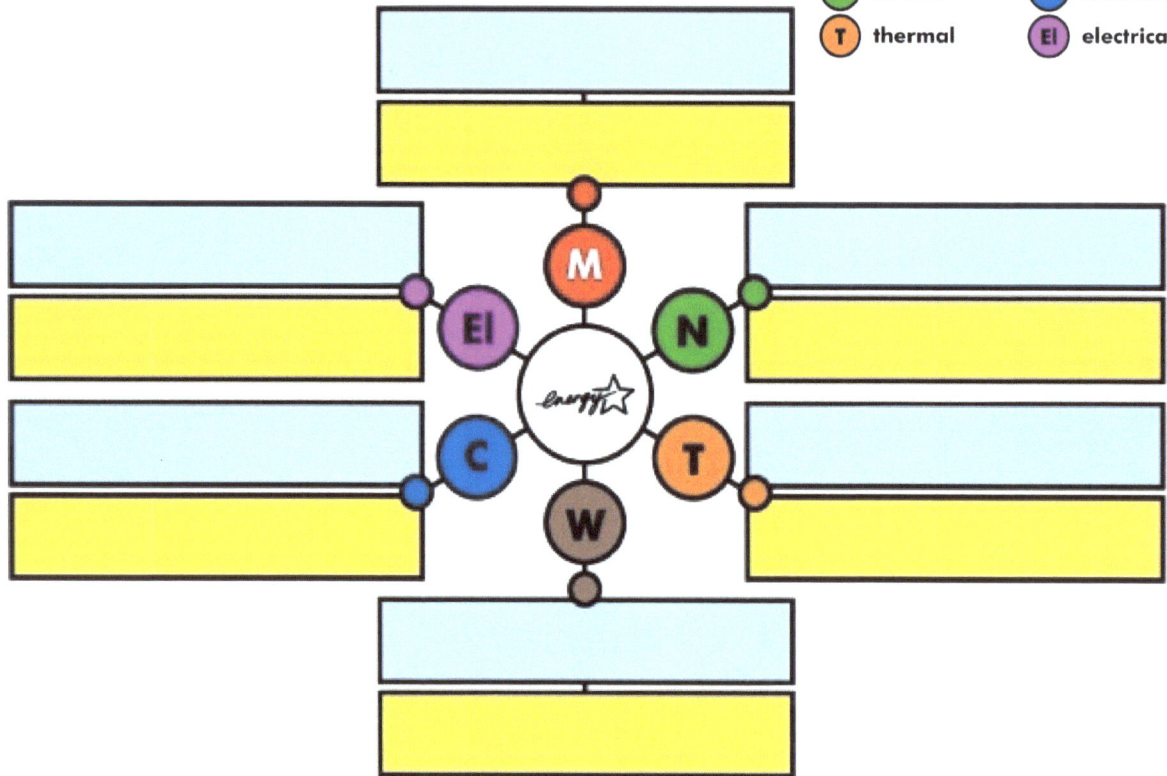

M	mechanical	**W** wave
N	nuclear	**C** chemical
T	thermal	**El** electrical

Directions

Write the description for the type of energy in the empty blue boxes in the energy web using the suggestions in blue boxes.

Write the examples of the type of energy in the empty yellow boxes by using the suggestions provided in the yellow boxes.

a moving bicycle, jumping frog, a hanging weight

static shocks, lightning power lines

food, matches

ice cream melting

nuclear power plants, the Sun's core

microwaves, visible light, ultraviolet radiation

energy inside atoms

energy of light and radiation

energy of the motion or position of an object

energy of electric charges

energy of heat

potential energy stored in bonds between atoms

Energy transfer is what makes all work possible.

Batteries don't make energy, they convert chemical energy into electrical energy. The metals inside batteries react with an acid or with a base that release electrons that can flow through a wire and make something work.

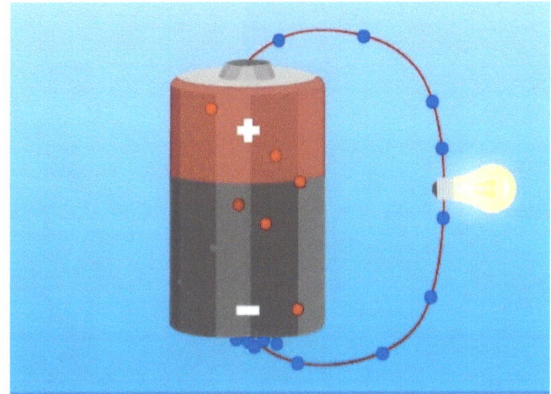

● Chemical energy
● Electrical energy

Photosynthesis is an example of energy transfer. The electro magnetic energy of sunlight is a form of wave energy that is transferred into chemical energy during the process of photosynthesis.

When you sharpen a pencil, you body transfers the chemical energy, that is derived from the food you eat, into mechanical energy. The energy is transferred to your arm and the pencil sharpener.

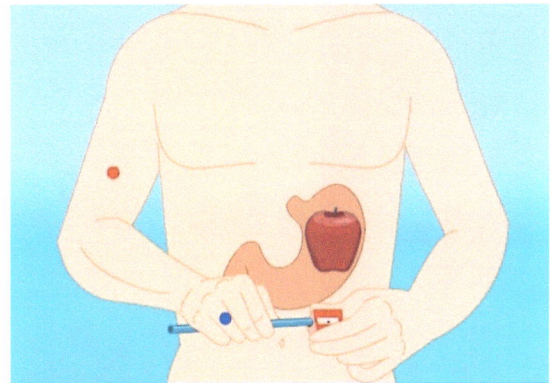

● Chemical energy
● Mechanical energy

www.onboardacademics.com

Identify the type of energy transfer.

A stove top transfers

_____

energy into

_____

energy.

A cyclist transfers

_____

energy into

_____

energy.

A light bulb transfers

_____

energy into

_____

energy.

electrical   chemical   wave

mechanical   electrical   thermal

Transferring Energy Quiz

1. Potential energy is stored energy. True or false?

2. A moving train is a good example of which of the following forms of energy.
   a. potential energy
   b. kinetic energy
   c. electrical Energy

3. _____ is the potential energy stored in the bonds between atoms.
   a. chemical energy
   b. nuclear energy
   c. thermal energy

4. Electromagnetic energy is the energy of electrical charges. True or false?

5. The energy stored inside atoms is _____.

6. The thermal energy of sunlight is converted into mechanical energy during the process of photosynthesis. True or false?

www.ingramcontent.com/pod-product-compliance
Lightning Source LLC
Chambersburg PA
CBHW052055190326
41519CB00002BA/236